A BIRD CAN FLY
by Douglas Florian

Greenwillow Books, New York

To Diane

Library of Congress Cataloging in Publication Data: Florian, Douglas. A bird can fly. Summary: Describes activities that a bird, beaver, ant, tortoise, monkey, camel, and a fish can and cannot do. 1. Animals—Miscellanea—Juvenile literature. [1. Animals—Miscellanea] I. Title. QL49.F59 591 79-24767 ISBN 0-688-80266-4 ISBN 0-688-84266-6 lib. bdg.

A bird can fly.

A bird can sing.

A bird can build a nest.

But a bird can't build a dam.
A beaver can build a dam.

A beaver can cut down trees
with its teeth.

A beaver can live underwater
for fifteen minutes.

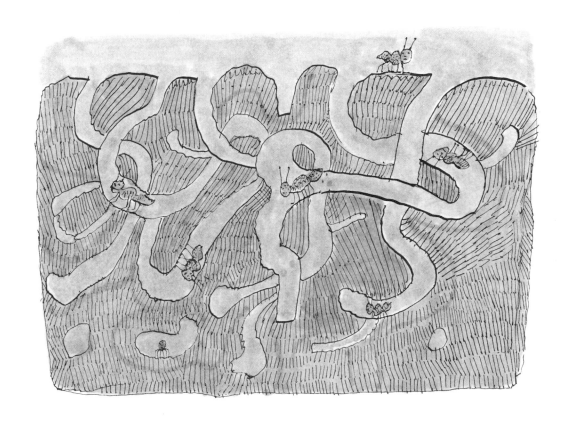

But a beaver can't live underground.
An ant can live underground.

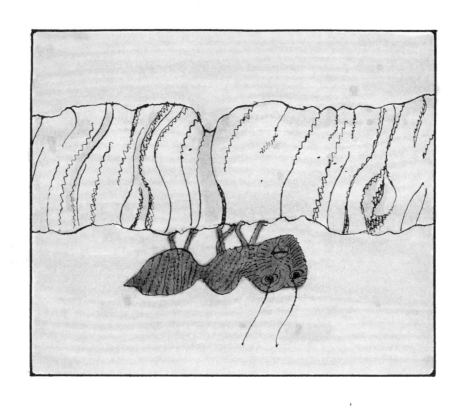

An ant can walk upside down
on a branch.

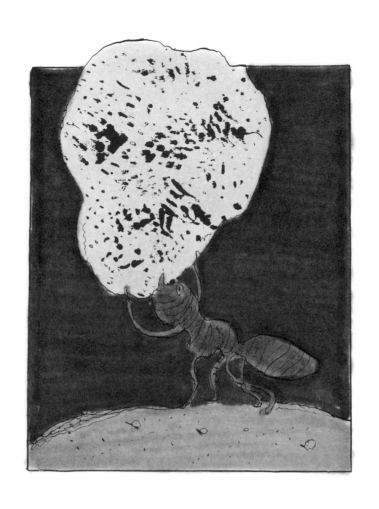

An ant can carry five times
its own weight.

But an ant can't carry a boy on its back.
A tortoise can carry a boy on its back.

A tortoise can live in a shell.

A tortoise can pull its tail
into its shell.

But a tortoise can't hang by its tail.
A monkey can hang by its tail.

A monkey can live in a tree.

A monkey can travel
through the jungle.

But a monkey can't travel
across the desert.
A camel can travel across the desert.

A camel can store food in its body.

A camel can drink twenty gallons of water.
But a camel can't breathe underwater.

A fish can breathe underwater.

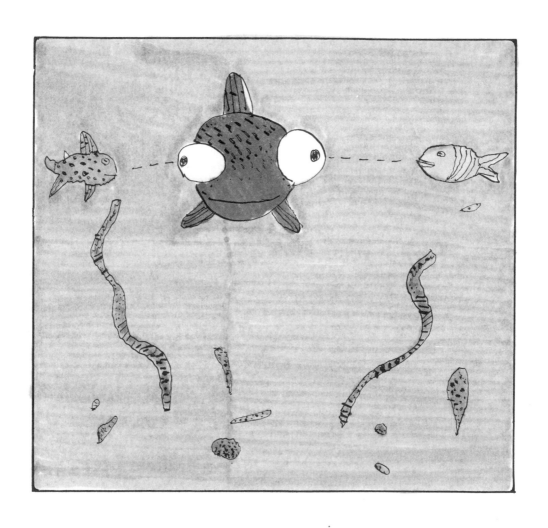

A fish can see things on both sides
of its head at the same time.

A fish can swim great distances.

But a fish can't fly.
A bird can fly.